Regulatory Primer for 2nd and Chief Engineers

Covering Current & New Regulations to 1 September 2013

Witherby Seamanship International
A Division of Witherby Publishing Group Ltd

4 Dunlop Square, Livingston, Edinburgh, EH54 8SB, Scotland, UK

Tel No: +44(0)1506 463 227 - Fax No: +44(0)1506 468 999
Email: info@emailws.com - Web: www.witherbyseamanship.com

First Published 2010
Second Edition Published 2011
Third Edition Published 2012
Fourth Edition Published 2013
Reprinted 2013

ISBN: 978-1-85609-615-7
eBook ISBN: 978-1-85609-616-4

© Witherby Publishing Group Ltd, 2010-2013.

British Library Cataloguing in Publication Data
A catalogue record for this book is available from the British Library.

Notice of Terms of Use

All rights reserved. No part of this publication may be reproduced, stored in a retrieval system, or transmitted in any form or by any means, electronic, mechanical, photocopying, recording or otherwise, without the prior permission of the publisher and copyright owner. While the principles discussed and the details given in this book are the product of careful study, the author and publisher cannot in any way guarantee the suitability of recommendations made in this book for individual problems, and they shall not be under any legal liability of any kind in respect of or arising out of the form or contents of this book or any error therein, or the reliance of any person thereon.

Printed and bound in Great Britain by Charlesworth Press, Wakefield

Published by

Witherby Publishing Group Ltd
4 Dunlop Square, Livingston
Edinburgh, EH54 8SB
Scotland, UK

Tel No: +44(0)1506 463 227
Fax No:+44(0)1506 468 999

Email: info@emailws.com
Web: www.witherbys.com

CONTENTS

Regulatory Primer for 2nd and Chief Engineers Covering Current & New Regulations

1 Fuel 1

2 Regulatory Related Questions 8

3 New Builds 32

4 Questions on Forthcoming Regulations Still to be Adopted 36

Regulatory Primer for 2nd and Chief Engineers Covering Current & New Regulations

1. Fuel

Que. **Bunkering**
From 1st January 2011, what now has to be carried and displayed during bunkering operations?

Ans. Material safety data sheets (MSDS) will need to be provided for ships carrying oil or oil fuel, prior to the loading of such oil as cargo in bulk or bunkering of oil fuel.

Ref: SOLAS Ch VI, Reg 5-1

Que. **Oil Tankers – Steam Driven Cargo Pumps**
What are the potential hazards on oil tankers with steam-driven cargo pumps that are operating in a port located in an ECA when switching from HFO to MGO?

Ans. The main safety concern is associated with switching from heavy fuel oil to marine gas oil in ships with boilers, which could increase the risk of furnace explosion caused by flame failure. The increased risk is a result of the temperature created in the boiler furnace during operation and the properties of the marine gas oil. A number of the recommended modifications when switching fuels in boilers need to be approved by class. Therefore, oil tankers that use the boilers to operate steam-driven cargo pumps for the discharge of cargo would need to allow the boilers to cool before changing the type of fuel.

Que. **ECAs**
What is an ECA?

Ans. This is the abbreviation for an Emission Control Area (ECA). An example would be the Baltic sea that was recognized as an ECA in May 2006 and the North sea in November 2007.

Que. **Low-Sulphur Fuels and ECAs**
What are the requirements for low-sulphur fuels in NW Europe?

Ans. Amendments to Marpol Annex VI in 2008, set a reduction of sulphur emissions in ECA's (emission control areas), these areas include the Baltic Sea, North Sea and English Channel. Ships in these areas will need to use fuels with a sulphur content of 1.0% from July 2010 and 0.1% sulphur by 2015.

Note:
Ships sailing in the rest of the world are working to a sulphur limit that is currently 4.5%. This will reduce to 3.5% in 2012, and then, pending a review of low sulphur fuel availability, 0.5% in 2020.

Que. **ECAs**
Your ship is going to sail to an area that is an ECA, what must you check when calculating the bunker requirement?

Ans. That before the ship enters the ECA, that she has sufficient fuel of 1.0% sulphur content already on board.

Que. **Monitoring and Enforcing the Use of Low Sulphur Fuels in the EU**
How will authorities monitor that ships are using low sulphur fuels in the EU?

Ans. Confirmation that the sulphur content of fuel oil is within the limit is to be obtained from the bunker delivery note provision of a shipboard installation that allows a fuel oil changeover logbook entry that confirms the fuel oil changeover prior to entry into a SO_x Emission Control Area (SECA) is required port state control inspections within the scope of IMO legislation and the framework of the Paris MOU.

Control of the sulphur content of marine diesel oil used at berth (EU legislation) is performed by the Environment Ministry in co-operation with the seaport police.

Que. **Bunker Delivery Note**
What must a bunker delivery note now show and how long must it be kept?

Ans. The sulphur content of each parcel of oil fuel intended for use on board ship will also require to be documented by means of a "Bunker Delivery Note" which must be kept on board for a period of 3 years after delivery of the fuel.

Que. Emissions – NO_x
What are NO_x?

Ans. *NO_x is a generic term for mono-nitrogen oxides (NO and NO_2). These oxides are produced during combustion, especially combustion at high temperatures.*

At ambient temperatures, the oxygen and nitrogen gases in air will not react with each other. In an internal combustion engine, combustion of a mixture of air and fuel produces combustion temperatures high enough to drive endothermic reactions between atmospheric nitrogen and oxygen in the flame, yielding various oxides of nitrogen. In areas of high motor vehicle traffic, such as in large cities, the amount of nitrogen oxides emitted into the atmosphere can be quite significant.

In the presence of excess oxygen (O_2), nitric oxide (NO) will be converted to nitrogen dioxide (NO_2), with the time required dependent on the concentration in air.

Que. NO_x Tier III
When are the MARPOL Annex VI NO_x Tier III standards expected to enter into force for new build ships?

Ans. It was expected that the NO_x (Nitrogen oxide) Tier III standards would enter into force on 1st January 2016, but at the IMO Marine Environmental Protection Committee (MEPC 65) in 2013, a delay until 2021 was agreed with lack of available technology being cited. For this amendment to be enforced it will need to be adopted at MEPC 66 in 2014.

Que. Emissions – SO_x
What are SO_x?

Ans. ***Sulphur oxide (SO_x)*** *refers to one or more of the following:*
Lower sulphur oxides (S_nO, S_7O_2 and S_6O_2)
Sulphur monoxide (SO)
Sulphur dioxide (SO_2)
Sulphur trioxide (SO_3)

Que. **Emissions – EU Ports**
What are the low sulphur requirements for ships in EU member ports if anchored or berthed at a port for more than two hours?

Ans. The EU directive on low sulphur fuels requires that ship's spending more than two hours either anchored or berthed in port must switch to using a marine gas oil that has a sulphur content of 0.1% or less.

Note:
Ref: Directive 2005/33/EC is intended to reduce shipping emissions of sulphur dioxide and particulates around coastal and port areas to prevent damage to the environment, human health and property, and thus reduce acid rainfall.

Que. **Exhaust Gas Cleaning**
What are exhaust gas cleaning systems?

Ans. They are an engineering solution that provides an alternative to low sulphur fuels and can reduce sulphur emissions down to 0.1%. Removing sulphur from the engine exhaust will allow a ship to continue to use a more regular, higher sulphur content and thus cheaper marine fuel oil.

Note:
*Exhaust gas cleaning systems are reputed to cost up to $3m USD.

Que. SO_x
What are the chemical groupings that are commonly referred to as SO_x?

Ans. Lower sulphur oxides (S_nO, S_7O_2 and S_6O_2)
Sulphur monoxide (SO)
Sulphur dioxide (SO_2)
Sulphur trioxide (SO_3).

Que. **Greenhouse Gases (GHGs)**
What do you understand by the phrase: *'Common But Differential Responsibilities'*?

Ans. The principle of Common but Differentiated Responsibility (CBDR) emerged as a principle from the 1992 Rio Earth Summit.

CBDR has two aspects. The first is the common responsibility, which arises from the concept of common heritage and common concern of humankind, and reflects the duty of States of equally sharing the burden of environmental protection for common resources; the second is the differentiated responsibility, which addresses substantive equality: unequal material, social and economic situations across States; different historical contributions to global environmental problems; and financial, technological and structural capacity to tackle those global problems. In this sense the principle establishes a conceptual framework for an equitable allocation of the costs of global environmental protection.

Que. **MARPOL Amendments – Revised Annex VI**
What were the recent amendments to MARPOL Annex VI that entered in to force on 1 July 2010?

Ans. The limits applicable in Sulphur Emission Control Areas (SECAs) were reduced to 1.00% on 1 July 2010 (from 1.50 %); being further reduced to 0.10 %, effective from 1 January 2015.

Que. **Sulphur Limits**
What were the changes in sulphur limits globally after 1st Jan 2012?

Ans. Reduction in sulphur limit to 3.5% globally for bunker fuels. (MARPOL Annex VI - Ch 3 Reg 14)

Note:
The revised Annex VI allows for an Emission Control Area to be designated for SO_x and particulate matter, or NO_x, or all three types of emissions from ships, subject to a proposal from a Party or Parties to the Annex, which would be considered for adoption by the Organization, if supported by a demonstrated need to prevent, reduce and control one or all three of those emissions from ships.

Que. **Fuel Switching**

The switching of fuel when entering an ECA presents a number of hazards, and is one of the leading causes of propulsion failures. In order to manage risk and improve safety, what measures can the ship take?

Ans.
- Consult engine and boiler manufacturers for fuel switching guidance;
- Consult manufacturers to determine if system modifications or additional safeguards are necessary for intended fuels;
- Develop detailed fuel switching procedures;
- Establish a fuel system inspection and maintenance schedule;
- Ensure system pressure and temperature alarms, flow indicators, filter differential pressure transmitters, etc., are all operational;
- Ensure system seals, gaskets, flanges, fittings, brackets and supports are maintained and in serviceable condition;
- Ensure a detailed system diagram is available;
- Conduct initial and periodic crew training;
- Exercise tight control when possible over the quality of the fuel oils received;
- Complete fuel switching well offshore prior to entering restricted waters or traffic lanes; and
- Test main propulsion machinery, ahead and astern, while on marine distillates.

Ref: USCG Marine Safety Alert regarding 'Fuel Switching Safety'
Issued: July 2011
http://tinyurl.com/9d7s88x

Que. **Reducing Ship Emissions**

What are the 3 different methods that can be used to reduce a ships emissions from its fuel oil?

Ans.
- Through reducing the amount of sulphur in the fuel before it is used
- using different abatement technologies to remove SO_x, NO_x and particulate matter after combustion but before emission
- use of a different fuel that naturally contains fewer polluting constituents and has lower emissions of CO_2 as well as SO_x, NO_x and PM.

Que. Abatement Technologies

Shipboard emissions can be reduced by the use of abatement technologies, what are abatement technologies?

Ans. Abatement technologies in a shipboard context describe exhaust gas cleaning equipment that provide a reduction in sulphur emissions from ships. This equipment removes SO_x, NO_x and particulate matter after combustion but before emission.

Que. Cold Ironing

What is 'cold ironing' (also called Alternative Maritime Power, AMP)?

Ans. During cold ironing, a ship at berth obtains its electricity for cargo handling and hotel requirements from a shore-based source via a high-voltage shore connection (HVSC). This method of saving onboard fuel is of particular value to certain types of ship that have high power requirements and spend a relatively long time at berth on a regular basis such as tankers, cruise ships, container ships and reefers.

Pollution concerns have made this technique an attractive option in some ports since it has an immediate impact on ship emissions. The use of cold ironing undoubtedly improves the air quality in and around ports by removing SO_x, NO_x and PM from the stack emissions. However, it does not necessarily have an overall positive impact on the total CO_2 emissions of fuel generation unless the electricity is produced from a non-fossil fuel source, such as nuclear or hydro.

2. Regulatory Related Questions

Que. **IMO Restructuring**
The IMO are undertaken a restructuring programme to improve effectiveness and reduce costs, do you know what it will affect?

Ans. The organisational restructuring will affect the Sub-committees which will be reorganised, renamed and reduced from 9 to 7. They will now comprise the following:

- Carriage of Cargoes and Containers (CCC)
- Pollution Prevention and Response (PPR)
- Implementation of IMO Instruments (III)
- Human Element, Training and Watchkeeping (HTW)
- Navigation, Communications and Search and Rescue (NCSR)
- Ship Design and Construction (SDC)
- Ship Systems and Equipment (SSE).

Que. **Ballast Water Systems**
What do you know about the BWM Convention coming in to force?

Ans. The International Convention for the Control and Management of Ships' Ballast Water and Sediments 2004 (BWM Convention 2004) will come in to force 12 months after ratification by 30 States, representing 35 per cent of world merchant shipping tonnage.

(July 2013 it has been ratified by 37 countries representing 30.32% of world merchant shipping tonnage).

Que. **Vessel General Permit**
What is the Vessel General Permit (VGP) and when does the existing coverage expire?

Ans. The VGP is required by Environmental Protection Agency (EPA) for ship operators to permit discharges incidental to the normal operation of a commercial vessel. The legislation issued by the US EPA is intended to eliminate all pollutants, including invasive species, from US inland and territorial waters and incorporates the US Coast Guard's mandatory ballast water management and exchange standards.

The initial VGP entered into force in 2009 and will expire on 19th December 2013. Ship operators are required to apply for coverage of their vessels under the VGP for ships:

- Greater than 79 feet in length
- greater than 300 GT
- with the capacity to hold or discharge more than 8 m^3 of ballast water.

The latest revision is largely aligned to the USCG ballast water rules where they adhere to the IMO's BWM Convention standards.

Que. **ECAs**
On the 1st August 2012, the 'North American Emission Control Area' came in to effect. What area does this ECA actually cover?

Ans. It comprises of the sea area located 200 nautical miles from the Atlantic, Gulf and Pacific coasts except where this impacts on the territorial waters of other States.

Que. **Arc Flash**
A recent 'M' Notice covered the hazards of 'Arc Flash' associated with high and low voltage equipment, what are these hazards?

Ans. This MGN highlights potential hazards of arc flash associated with high and low voltage electrical equipment onboard vessels.

Key Points:

- The best approach for electrical safety to prevent an arc flash incident is to only perform work on de-energised equipment that has been placed into an electrically safe condition.
- After control measures to reduce the risk of an arc flash have been investigated/implemented, protective clothing and PPE requirements should be carefully selected.

- One of the major hazards associated with an arc flash is burn injury from the exposure to the thermal energy from an arc flash.

Ref MGN 452

Que. **Inventory of Hazardous Materials (IHM)**
In 2011, the IMO adopted Resolution MEPC.197(62) Guidelines for the Development of the Inventory of Hazardous Materials (IHM). The guidelines provide recommendations for ships of 500 gt and above for the hazardous waste inventory required for the issue of the International Certificate on Inventory of Hazardous Materials, as required by Regulation 5 of the Hong Kong International Convention.

The Guidelines will remain voluntary until the Convention is ratified and enters into force, making them mandatory.

What is contained in an Inventory of Hazardous Materials (IHM)?

Ans. The IHM is recommended to be structured in 3 Parts:

- Part I - Materials contained in ship structure or equipment
- Part II - Operationally generated wastes
- Part III - Stores.

Part I - Materials contained in ship structure or equipment

Table A includes the following substances and where they may be found on board:

- Asbestos
- polychlorinated biphenyls (PCB)
- ozone depleting substances
- anti-fouling systems containing organotin compounds as a biocide.

Table B is primarily aimed at new ships, but encourages existing ships to identify substances to support the recycling process.

It provides an indicative list of substances (and compounds), specifically:

- Cadmium
- hexavalent chromium
- mercury
- lead
- polybrominated biphenyls (PBBs)
- polybrominated diphenyl ethers (PBDE)
- polychlorinated naphthalenes
- radioactive substances
- certain shortchain chlorinated paraffins.

These materials need not be listed when they are used in general construction and are inherent in metals or metal alloys.

Part II - Operationally generated wastes

Table C lists potentially hazardous solids, liquids and gases including lubricants, hydraulic oils, solvents, refrigerants, antifreeze, propane, oxygen, garbage, incinerator ash, cargo residues, etc. The table identifies whether they should be listed under Part II or Part III (stores).

Part III - Stores

Table D describes regular consumable goods, potentially containing hazardous materials, that are not ship-specific operational equipment, eg electrical goods, office equipment, personal computers, light bulbs, TVs, etc.

Any spare parts containing materials listed in Tables A or B are required to be listed in Part III.

A standard format for Tables A and B is provided in Appendix 2 of the Guidelines, to indicate the material, location and quantity on board, with space for any additional information.

Que. **Designated Person Ashore (DPA)**
What is the role of the Designated Person Ashore (DPA)?

Ans. "To ensure the safe operation of each ship and to provide a link between the Company and those on board, every Company, as appropriate, should designate a person or persons ashore having direct access to the highest level of management. The responsibility and authority of the designated person or persons should include monitoring the safety and pollution-prevention aspects of the operation of each ship and ensuring that adequate resources and shore-based support are applied, as required."

Ref: ISM Code section 4

If you break this down further the four specific areas of responsibility (or function) that the DPA has, are as follows:

1. provide a link between the Company and those on board
2. having direct access to the highest level of management
3. monitoring the safety and pollution-prevention aspects of the operation of each ship
4. that adequate resources and shore-based support are applied

Que. **Annex VI of MARPOL**
Why did ships such as LNG Carriers receive an exemption to parts of Annex VI of MARPOL?

Ans. In October 2010 a revision of MARPOL Annex VI exempted steamships from the fuel sulphur limits associated with the North American [or the United States Caribbean] Emission Control Areas (ECAs), which were to be adopted in July 2012.

The exemption recognizes the risks associated with switching from residual fuel to distillate fuel in boilers.

Note:
Emission Control Areas (ECAs) came into effect through MARPOL Annex VI, Reg 14 for three separate areas: the Baltic Sea, the North Sea and North America. The current

sulphur limit for fuel used in these areas reduced to 1.00% as of the 1st July 2010, and will be further reduced to 0.10% on 1st January 2015.

Merchant ships have a number of options to comply with these fuel sulphur limits when they enter a designated ECA. They may be able to switch to a low sulphur residual fuel if available or they may switch to a marine distillate fuel, as would be the case to meet the 0.10% standard from 2015, or they could use an alternative method such as an exhaust gas cleaning system (EGCS) that achieves the required SOx and particulate matter reductions.

Most ships that run on residual fuel are designed to, or can be modified to, operate on distillate fuel. However, there remains a very small number of ships that are propelled by steam boilers where the different combustion techniques that are used presents a substantial and costly challenge to comply with the fuel sulphur content requirements for ECAs.

An independent report to the European Commission in 2009 demonstrated that up to 260 LNG carriers would need to adapt their boilers to ultra-low sulphur fuels to meet these changes if they wish to continue trading in these ECAs, with the estimated cost to adapt a marine boiler on an LNG carrier being suggested at around $2.1M. This is before considering where an operator could obtain the necessary expertise to conduct the required modifications; there are reports that such changes could not feasibly be completed for the 260 LNG Carriers before 2020.

While it is possible to modify a marine boiler on a steamship to operate on ultra-low sulphur marine distillate, the main propulsion boilers are often of a rare design that is no longer supported by manufacturers. Therefore, the ability to switch fuel types as a compliance option is less viable for these ships. Without the correct modifications and clear instructions on the procedure to switch fuel types, these ships present an increased risk of boiler explosion. When the age of the vessels is taken into account, it does suggest that an alternative approach may be required for steamships operating in ECAs.

Que. **Do You Know the Phase-in Requirement for Ballast Water Systems on Your Last Ship?**

Ans. **Phase-in deadline for fitting ballast water systems**

Existing tonnage delivered before 2009)		Newbuildings (keels laid 2009 onwards)	
Ship type	Year	Ship type	Year
Existing vessels with ballast capacity between 1,500 and 5,000 m^3	2014	Keels laid 2010 onwards, ballast capacity less than 5,000 m^3	2011
All other existing vessels (less than 1,500 cu m or higher than 5,000 m^3)	2016	Keels laid between 2009 and 2012, ballast capacity over 5,000 m^3	2016
		Keels laid 2012 onwards, ballast capacity over 5,000 m^3	Upon delivery

Note:

In state inspection terms, the deadlines are usually for the first scheduled survey after the date for the deadline

The purchase and installation costs of a ballast water treatment plant could cost between $500,000 and $2 m per vessel.

Que. **Piracy**

Where would you find information on measures and practices that you could implement onboard your ship to deter the likelihood of a Piracy attack?

Ans. There is an IMO circular on '*Piracy and armed robbery against ships in waters off the coast of Somalia*' which includes Best Management Practices to Deter Piracy in the Gulf of Aden and off the Coast of Somalia. These have been developed by industry organisations.

The Best Management Practices (BMP4) is a little blue book that was published in Summer 2011. This is a 96 page booklet that is available free of charge from chart agents and the industry trade associations.

Note:
BMP4

The purpose of the Industry Best Management Practices (BMP) is to assist ships to avoid, deter or delay piracy attacks off the coast of Somalia, including the Gulf of Aden (GoA) and the Arabian Sea area. Experience, supported by data collected by Naval forces, shows that the application of the recommendations contained within this booklet can and will make a significant difference in preventing a ship becoming a victim of piracy.

For the purposes of the BMP the term 'piracy' includes all acts of violence against ships, her crew and cargo. This includes armed robbery and attempts to board and take control of the ship, wherever this may take place.

Where possible, this booklet should be read with reference to the Maritime Security Centre – Horn of Africa website (www.mschoa.org), which provides additional and updating advice.

Que. **Green Passport**
Do you know what a ship's 'Green Passport' is?

Ans. The Ship Recycling Convention introduces the concept of a ship's 'Green Passport', which is essentially an inventory of materials present in a ship's structure, systems and equipment that may be hazardous to health or the environment. This is kept up-to-date for the service life of the ship. Prior to breaking, details of any further hazards and waste material are added, which will help the recycling yard to develop a safer and more environmentally acceptable plan.

Green Passports, which are voluntary at the current time, are typically verified by Classification Societies who now provide an approval and verification service for newbuildings and existing ships.

For ISO 14001 certified companies, a Green Passport for each ship assists in demonstrating best practice in managing tonnage in an environmentally responsible manner.

Que. **Goal-Based Construction Standards**
What are goal-based construction standards?

Ans. Goal-based construction standards are to be enshrined into five tiers. The standards will be set within the top three tiers and the detailed requirements in the fourth and fifth. They are not about setting standards for individual ships, but govern the development of the rules and regulations that impact on ship design.

In its very simplest form, tier one is the ultimate goal: a ship designed to have a safe operating life of 25 years. The second tier consists of the functional requirements on how that is achieved. This could consist of objectives such as sufficient structural strength, power generation, seakeeping performance, or any one of a range of other needs for the vessel to remain safe.

The third tier is the verification of compliance, by the IMO, that the detailed requirements in the final two tiers meet these stated objectives in tier two. It is currently requesting that verification documents be submitted by the end of 2012 to allow goal-based standards to be adopted and in force by 2015.

Note:
Goal Based Standards for Shipbuilding. Guidelines for Verification (non-mandatory).

Guidelines for Information to be included in a ship construction file (non mandatory)
This will apply to ships:
1. building contract on or after 1 January 2015; or
2. keel lay on or after 1 January 2016; or
3. delivery on or after 1 January 2019.

Que. **The Human Element**
What do you understand by the use of the phrase 'The Human Element'?

Ans. There is a 2010 publication sponsored by the MCA. It gives an insight as to why people take risks often with "dreadful consequences" and that decisions are a "*trade-off between: available information and available time*".
There is a strong view that what the shipping industry really lacks is training in dealing with the human factor. This guide covers stress, the dos and don'ts of fatigue, and provides valuable advice

on working with others, and communications. It also tells masters how to take control of a ship simply and effectively.

Note:
The book's inspiration was a train crash in 2002 that killed seven and injured 76. After the incident the authors wrote a 200-page guide to human factors in the rail industry. In 2008 they did a three-month feasibility study on the shipping industry. This book is the result dealing with the IMO, company boardrooms to officers and crew.

Que. **Noise levels**
What is the Code on Noise Levels on Board Ships?

Ans. The Code is intended to provide standards to prevent the occurrence of potentially hazardous noise levels on board ships and to provide standards for an acceptable environment for seafarers. These standards were developed to address passenger and cargo ships.

The Code is intended to provide the basis for a design standard, with compliance based on the satisfactory conclusion of sea trials that result in issuance of a Noise Survey Report.

Ongoing operational compliance is predicated on the crew being trained in the principles of personal protection and maintenance of mitigation measures. These would be enforced under the dynamic processes and practices put in place under SOLAS chapter IX.

The Code enters into force on 1st July 2014 and will apply to all new ships greater than 1,600 gt.

Que. **New Rules for Hours of Rest**
The revisions to the STCW Convention will come into force in 2012. What are the new requirements for hours of rest?

Ans.
- Rest periods of not less than 10 hours in any 24 hour period and 77 hours in any seven day period
- No more than two rest periods, one of which must be at least six hours; and
- Intervals of no more than 14 hours between rest periods.

Exceptions in cases of emergency and overriding operational conditions:

- Reduction in minimum rest hours to 70 in a seven day period, but for a maximum of two weeks and a gap of twice the period of exception, before there is any further exception
- Increase in rest periods from two to three – one of at least 6 hours and the others no less than one hour
- Interval between rest periods no more than 14 hours; and
- Only applicable for two days in any seven day period.

Que. **STCW**
When did the recent Manila amendments to the STCW Convention and Code come in to force?

Ans. On the 1st January 2012

Note:
These are the first updates to the STCW Convention and Code since 1995. For seafarers who started their seagoing service before 1 July 2013, the existing arrangements will apply until 1 January 2017. For more information go to: http://tinyurl.com/stcw2010-summary

Que. **STCW Manila Amendments**
By what date are approved courses to meet and be certified in accordance with the Manila amendments to the STCW Convention and Code?

Ans. 1st July 2013 - by this date governments will need to submit compliance documents to remain on "white list".

Que. **Port State Control**
What do you understand by the term 'Port State Control'?

Ans. Port State Control is a check on visiting foreign ships to see that they comply with international rules on safety, pollution prevention and seafarers living and working conditions. It is a means of enforcing compliance where the owner and flag State have failed in their responsibility to implement or ensure compliance. The port

Regulatory Primer for 2nd and Chief Engineers

Que. **Port State Control**

State can require defects to be put right, and detain the ship for this purpose if necessary. It is therefore also a port State's defence against visiting substandard shipping.

On the 1st January 2011 the New Inspection Regime (NIR) of the Paris MoU on Port State entered into force. What are the key points of this New Inspection Regime (NIR)?

Ans. Key points of the NIR

- Ships will be targeted based on a risk profile that considers two elements - Ship Risk Profile and Company Performance
- the Ship Risk Profile (SRP) classifies ships into one of three categories: Low Risk Ships, Standard Risk Ships and High Risk Ships. While the SRP assesses type and age of ship, number of previous deficiencies and detentions, performance of the flag of the ship, and the performance of the recognised organisation(s), the it also considers a Company's performance and takes into account the detention and deficiency history of all ships in a (ISM) company's fleet in the Paris MoU area in the last 36 months
- high risk ships will be due inspections every 5 - 6 months, standard risk ships every 10 - 12 months and low-risk ships every 24 - 36 months
- additional inspections may be carried out between these intervals, for reasons such as reports from pilots, collisions, groundings, etc
- the type of inspection will depend on the ship's risk profile; the minimum for a high risk ship will be an expanded inspection
- ships requiring an expanded inspection must give 72 hours notice prior to arrival.

Note:
To maintain low risk status, a vessel must have no more than 5 deficiencies during any one inspection and no detention recorded in the preceding 3 years.

Que. **Port State Control**
From 1 January 2011, what is the 'Inspection Frequency' under the New Inspection Regime (NIR) of the Paris MoU on Port State Control?

Ans. Port State Inspections conducted under the previous regime will count. Therefore, for example, if a ship was inspected on 21st October 2010 and, under the new regime, is designated a Standard Risk ship, the window for inspection will open on 21st August 2011 (ie 10 months after last inspection) and the ship will be Priority II. The ship will become Priority I from the 21st October 2011 (ie 12 months since the last inspection) and must be inspected.

Note:
The new system is more prescriptive in that, depending on the risk profile of a ship, it will be known when the next periodic inspection is due. Therefore, for a Standard Risk ship, once an inspection has taken place the ship could expect an inspection free period of at least 10 months. The ship could be inspected within the next 2 months but will know that after 12 months it will be inspected at the next Paris MoU port.

Que. **Port State Control**
Where would you find guidance when preparing for a Port State Control Inspection?

Ans. A number of industry bodies (e.g. Classification Societies, P&I Clubs, etc) have produced checklists and guidance relating to preparation for PSC inspection. Companies may find these useful in ensuring that their vessels are suitably prepared for PSC inspections.

Que. **Port State Control**
Who determines if a ship is to be detained after a Port State Control Inspection? how is that decision arrived at and who should the Master notify?

Ans. The Port State Control Officer (PSCO) will exercise his professional judgement in determining whether to detain the ship until the deficiencies are corrected or to allow it to sail with certain deficiencies without unreasonable danger to the safety, health, or the environment, having regard to the particular circumstances of the intended voyage.

The ship's operator should report any PSC detentions at the first opportunity to their flag state, the issuing body of the affected certificate and their ISM issuing body.

Que. Port State Control

Your ship has been detained in port by Port State Control and the ship's operator has notified the ship's flag State, what would you expect to be the typical response from a flag State?

Ans. This will vary based on the specific flag State and the various issuing and certifying authorities concerned, but generally involves the ship's operator preparing a report of the deficiencies found, in conjunction with a root cause analysis and a note of what subsequent corrective actions have been taken to prevent reoccurrence on the company's ships.

Typically, the flag State will examine these reports and determine if any further action is appropriate. Such measures may include an additional audit of the ISM Safety Management Certificate (SMC) and/or the ISM Document of Compliance (DOC) may be required to verify that the Safety Management System is operating effectively.

If the detention occurs within the survey window for a related annual, periodical or intermediate survey, the ship can expect that this survey will require to be completed before the ship sails.

However, if the detention occurs outside of the related survey windows, the surveyor, after addressing the PSC deficiencies, will conduct a general examination of the ship, to determine if any additional surveys are required.

Depending on the flag State, further sanctions may be imposed for ships that have experienced multiple detentions in a specific period. As an example, the following is how the Bahamas Maritime Authority responds in such cases:

- If a ship has been justifiably detained twice within a period of 24 months, an immediate additional ISM SMC audit to the extent of initial audit will be required to ascertain the effectiveness of the Safety Management System on board. Furthermore, an additional ISM DOC audit to the extent of annual audit will be required not later than 30 days from the date of the detention
- if a ship has been justifiably detained three times within a period of 24 months, all statutory certificates will be suspended. In order to reinstate these, renewal surveys should be carried out with no outstanding items or recommendations to the extent possible, and an additional ISM SMC audit to the extent of initial audit will be required. Furthermore, an additional ISM DOC audit to the extent of initial audit will be required not later than 30 days from the date of the detention
- furthermore, if a ship is justifiably detained for a fourth time within a period of 24 months, the vessel will be deleted from the Bahamas Registry. The DOC of the Manager, issued on behalf of the Administration, will be re-examined
- the imposition of a flag State Detention will have the same effect, for the purposes of this process, as a port State Detention
- the BMA will also decide the scope and extent of additional inspection or survey or additional audits of shipboard and shore based Safety Management Systems of a Company, when a significant proportion of the Company fleet is justifiably detained by PSC.

Que. **Port State Control**

Substandard operators identified during PSC inspections will be 'named and shamed' via a new online register, while shipowners with strong safety records will be given good public visibility. Under this regime, what is the 'Black List'?

Ans. This is a table showing the port state control performance of fleets whose detention ratio over a 3 year rolling period was above the average. The spectrum of the poor performers is ranked in 4 categories: medium risk, medium to high risk, high risk and very high risk.

Que. **Port State Control**
Substandard operators identified during PSC inspections will be 'named and shamed' via a new online register, while shipowners with strong safety records will be given good public visibility. Under this regime, what is the 'Grey List'?

Ans. Flags with an average performance are shown on the 'Grey List'. Their appearance on this list may act as an incentive to improve and move to the 'White List'. At the same time flags at the lower end of the 'Grey List' should be careful not to neglect control over their ships and risk ending up on the 'Black List' next year.

Que. **Port State Control**
Substandard operators identified during PSC inspections will be 'named and shamed' via a new online register, while shipowners with strong safety records will be given good public visibility. Under this regime, what is the 'White List'?

Ans. The 'White List' represents quality flags with a consistently low detention record.

Que. **Port State Control**
Under the New Inspection Regime (NIR) of the Paris MoU on Port State Control, what does an initial inspection consist of?

Ans. An initial inspection will consist of a visit on board the ship to:

- check the certificates and documents listed in Annex 10;
- check that the overall condition and hygiene of the ship including:
  - navigation bridge
  - accommodation and galley

- decks including forecastle
- cargo holds/area
- engine room

meets generally accepted international rules and standards

- verify, if it has not previously been done, whether any deficiencies found by an Authority at a previous inspection have been rectified in accordance with the time specified in the inspection report.

Que. **Accident Reporting (UK)**
What changes have been made to the MAIB accident reporting and investigation system?

Ans. The definition of *accident* has been amended and includes *marine casualties* and *marine incidents*. It also now covers *serious injuries*, the definition of which has been expanded to include:

a. any fracture, other than to a finger, thumb or toe;
b. any loss of a limb or part of a limb;
c. dislocation of the shoulder, hip, knee or spine;
d. loss of sight, whether temporary or permanent;
e. penetrating injury to the eye;
f. any injury to a person employed or carried in a ship which occurs on board or during access which results in incapacitation for more than three consecutive days excluding the day of the accident, or
g. any other injury leading to hypothermia, unconsciousness, requires resuscitation or requiring admittance to hospital or other medical facility as an in-patient for more than 24 hours.

The *Incident Report Form* has now been replaced by an *Accident Report Form. MGN 458 (M+F) - Accident Reporting and Investigation.*

Que. Ballast Water Treatment Systems

Of the many different types of new ballast water treatment systems that you will see onboard ships over the next few years, there are a number that are using Sodium Hypochlorite to treat the ballast water. Are there any concerns with this type of treatment plant?

Ans. Sodium hypochlorite is produced by the electrolysis of seawater within a hypochlorite generation cell, by the action of the seawater being forced to flow between two concentric titanium tubes that are connected to a DC (direct current) power supply.

Sodium hypochlorite solution is frequently injected into sea-chests on offshore units that are static such as oil production platforms, drilling rigs and FPSOs to combat the growth of marine organisms and algae that foul filters and seawater pipelines.

The electrolysis of seawater will produce hydrogen gas, but the quantities generated are insufficient under safe operating conditions to constitute a hazard. Hydrogen gas is extremely dangerous as it is highly flammable with its LEL 4.1% and UEL 74%.

The process of venting hydrogen from the ballast water is very important as, without the use of a hydrogen separator, it is possible for hydrogen to be introduced into the ballast water tanks where it could potentially reach a hydrogen-in-air concentration that enters the flammable range. In addition, a further hazard is that any leaks of hypochlorite mixing with acidic solutions would result in the immediate formation of chlorine (Cl_2) gas, which can be fatal.

Note:
Case Study – Operator error while hypochlorite dosing a sea-chest results in explosion

As discussed, sodium hypochlorite has been used for dosing sea-chests on offshore installations for many years and it is from that industry that we can learn further about the severity of the hazard when operating equipment that generates chlorine gas. This incident occurred on an FPSO that had been retrofitted with enlarged sea-chests to accommodate the substantial cooling needs of a petroleum processing plant.
To give an idea of the size of these sea-chests, the top mounted flange on the sea-chest was approximately 48" in diameter secured by approximately 32 bolt

positions with 32-36 mm securing nuts. A change-over of the sea-chest in use, from the starboard to the port, was conducted and the starboard sea-chest duly isolated. However, the sodium hypochlorite generator was left running dosing the static volume of water in the sea-chest.

The explosion, resulting from the build-up of chlorine gas in the sea-chest some time later, tore the sea-chest flange off and bent it in with the same apparent ease as if it was the polystyrene base beneath a pizza, shearing the majority of the securing bolts in the process.

The explosion was felt all around the FPSO and she was lucky not to have suffered a worse fate after receiving an explosion to such a vulnerable area beneath her waterline.

The manufacturers comment:
We contacted all the manufacturers of ballast treatment systems that use sodium hypochlorite. They confirmed that this concern had been addressed at IMOs Marine Environmental Protection Committee and, in its simplest form, the risk could be regarded as you would the radioactive risk on an item of electronic equipment, ie yes, the risk is there but the volumes are too small. These are new technologies and new treatment methodologies that seafarers are not experienced with and the very nature of these treatment additives emphasizes the caution that they should be operated with, and never outside of the parameters determined by the manufacturer.

Que. **Amendments to ISM Code**

What were the amendments to the ISM Code that entered force on the 1st July 2010?

Ans. In addition to amending the requirements for the renewal verification and certification of the Safety Management Certificate (SMC) and the Document of Compliance (DOC), greater emphasis is placed on companies defining procedures rather than making preparations for the facilitation of them.

The changes will have an impact on the content, auditing and approval of Safety Management Systems (SMS), and in some cases may require a more detailed, prescriptive and voluminous content than presently exists.

Note:
The Company must carry out internal safety audits on board and ashore at intervals not exceeding twelve months to verify whether safety and pollution prevention

activities comply with the safety management system. In exceptional circumstances, this interval may be exceeded by not more than three months.

The SMS must now be periodically reviewed by the Master. When establishing procedures for the implementation of corrective action, the procedures must now include measures intended to prevent recurrence.

Specific requirements for renewing certificates, similar to that provided for other SOLAS certificates now apply to the SMC.

Que. **2010 MARPOL Annex IV Sewage Amendments**
What were the recent changes?

Ans. On 1st August 2010, the amendments to MARPOL Annex IV came in to force for existing ships.

The original legislation entered into force on 1st August 2005 applying only to new ships on international voyages of 400 gross tons and above or certified to carry more than 15 persons.

Under the regulations, ships are prohibited from discharging sewage unless one of the following criteria is complied with:

- Uncomminuted or non-disinfected sewage can only be discharged at a distance more than twelve nautical miles from the nearest land
- comminuted and disinfected sewage can be discharged at a distance more than three nautical miles from the nearest land
- certified sewage treatment plant is in operation meeting regulation 9.1.1 of Annex 5.

This requires ships to be equipped with one of the following:
- A type approved sewage treatment plant
- an approved sewage communiting and disinfecting system
- a sewage holding tank.

Que. **Annex IV Special Area**
When did the Baltic Sea designated as a Special Area under MARPOL Annex IV (Sewage)?

Ans. The Baltic Sea was designated as an Annex IV Special area on 1st January 2013.

Que. **Fixed Carbon Dioxide Fire-extinguishing Systems**
What are the requirements that came in to force for all ships >500 t from 1st January 2010?

Ans. Fixed carbon dioxide fire-extinguishing systems for the protection of machinery spaces and cargo pumprooms are to be upgraded to comply with the provisions for control under the Fire Safety Systems Code. This includes two separate controls located inside a release box clearly identified for the particular space shall be provided to release the CO_2 and, upon release, an audible alarm shall be activated.

Ref: SOLAS Ch II-2 Fixed CO_2 System Upgrades

Que. **Oil Tankers – Single Hull Phase Out**
What are the phase out requirements for single hull oil tankers?

Ans. Regulation 13G of Annex I to MARPOL 73/78 came into force on 6th July 1995. It imposed requirements on existing crude oil tankers concerning double hulls, enhanced survey and scrapping, with the aim of either converting or scrapping of all single hull tankers by 2015. This Regulation is now embodied in Regulation 20 in the current consolidated edition of MARPOL.

Note:
Under the revised Regulation 13G of MARPOL Annex 1, the final phase-out dates were brought forward as follows:
Category 1 tankers (pre-MARPOL tankers) – from 2007 to 2005
Category 2 and 3 tankers (MARPOL and smaller tankers) – from 2015 to 2010.

The remaining timetable for the phasing out of single hull tankers is:

CATEGORY	DATE
Category 2: 20,000 dwt and above carrying • crude oil, fuel oil, heavy diesel oil or lubricating oil as cargo 30,000 dwt and above carrying • other oils, which do comply with the protectively located SBT requirements (MARPOL tankers) **Category 3:** 5,000 dwt and above but less • than Category 2	Anniversary date in 2009 for ships delivered in 1983 Anniversary date in 2010 for ships delivered in 1984 or later

In the case of certain Category 2 or 3 oil tankers fitted only with double bottoms or double sides extending the entire cargo tank length and not used for the carriage of oil, and tankers fitted with double hull spaces not meeting the minimum distance protection requirements which extend to the entire cargo tank length and are not used for the carriage of oil, the Flag State may, if satisfied that the ship complies with specific conditions, permit continued operation beyond a vessel's phase-out date where the ship was in service on 1st July 2001. However, this cannot extend beyond 25 years after delivery, and these vessels will only be able to trade in a very limited area.

Que. **LSA Calculations**

What is the average mass of a person under the recent LSA Code revisions to be used in calculations?

Ans. The average mass of a person, when determining the carrying capacity of cargo ship lifeboats (including free-fall lifeboats) and all rescue boats, is increased from 75 kg to 82.5 kg.

Note:
For free-fall lifeboats, the requirements for seats, seat arrangement and passage between seats have been revised.

Que. Passenger Ship Safety

Since the Costa Concordia grounding what recent SOLAS amendment is due to enter into force at the start of 2015, and what updates have recently been made to the interim guidelines for passenger ship safety?

Ans. SOLAS III/19 will require that a passenger muster is required *'prior to or immediately after'* departure. MSC 92 saw the following interim guideline amendments, and these will be issued as MSC.1/Circ.1446/Rev.2:

- Securing of heavy objects. This update provides guidelines on heavy objects found on board ships that should be:
    - Permanently secured (e.g. pianos, cash machines, treadmills)
    - secured when not in use (e.g. heavy chemical containers, trolleys, x-ray scanners)
    - secured in heavy weather (e.g. display items, repair/refurbishment equipment, temporary decorations)
- Harmonisation of bridge navigational procedures. It was recommended that wherever possible the procedures used on the bridges of ships within the same fleet or management group be consistent.
- Stowage and provision of lifejackets. This update suggests that lifejackets should be stowed near muster points or lifeboats so that passengers do not need to return to their cabins to collect lifejackets in the event of an emergency.
- Voyage planning. It is suggested that the importance of adhering to voyage plans be reinforced and also that voyage planning guidelines be strictly followed in case deviations from the plan are necessary.
- Passenger emergency instructions. It was suggested that videos be used to convey passenger emergency instructions and that emergency information cards are made available to passengers.

Que. MES
How frequently should the rotational deployment of Marine Evacuation Systems (MES) be carried out?

Ans. SOLAS III/20.8.2 requires that the rotational deployment cannot exceed 6 years. Servicing of MES should be carried out annually.

The Maritime and Coastguard Agency's Marine Guidance Note, MGN 463, provides clarification in regard to MES servicing and rotational deployment and the roles that different parties should play in the deployment and the fail criteria for the deployment.

Que. CFCs
Is the use of CFCs permissable onboard?

Ans. The use of CFCs (mainly utilized in air conditioning and refrigeration systems) is prohibited on ships flying the flag of a MARPOL VI signatory State.

Ref: Revised MARPOL Annex VI, Reg 12 Use of CFCs, entry in to force: 01/07/2010

Que. PSSA
When was the Saba Bank designated as a Particularly Sensitive Sea Area (PSSA)?

Ans. The Saba Bank, Caribbean Sea, was designated as a PSSA effective 5th October 2012.

Que. Asbestos
There was a SOLAS Regulation came in to force on the 1st January 2011 concerning asbestos. What was the regulation about?

Ans. It prohibits all new installations of asbestos on board ships.

3. New Builds

Que. **Machinery Space – New Build Chemical Tanker**
What are the requirements that entered force on 1st January 2009 for new build Chemical tankers >2000 t?

Ans. Any category A machinery space exceeding 500 m³ in volume must be provided with an approved type of fixed water-based or equivalent local application fire-fighting system, based on MSC/Circ.913 in addition to the required fixed fire extinguishing system.

Ref: IBC Code & MSC.219(82)

Que. **Passenger Ship – Flooding Detection System**
What were the requirements that entered force for new builds after 1st January 2009 carrying more than 36 passengers?

Ans. All tanks and watertight spaces located below the bulkhead deck are to be provided with a flooding detection system.

Ref: SOLAS Ch II-1 Flooding Detection

Que. **Vehicle Spaces – Drainage**
What are the requirements for drainage openings from closed vehicle spaces?

Ans. On new build cargo ships >500 gt from 1st January 2010, drainage openings from closed vehicle spaces, Ro-Ro spaces, or special category spaces that are protected by fixed pressure water-spraying systems are to be fitted with a nonoperational means to prevent blockage.

Ref: SOLAS Ch II-2, Reg 20.6 Drainage System Protection

Que. **Passenger Ships Fire Detection and Alarm**
What are the requirements on new build passenger ships carrying more than 12 passengers built after 1st January 2010?

Ans. Fixed fire detection and fire alarm systems are to be capable of remotely and individually identifying each detector and manually operated call point. Any section of fire detectors and manually operated call points shall not be situated in more than one main vertical zone.

Ref: FSS Code

Que. **Ventilation Ducts**
What are the requirements on all new build ships >500 t built after 1st July 2010?

Ans. Ducts are to be constructed of steel or equivalent material (as opposed to a non-combustible material). Short ducts (2 m) need not comply provided the ducts are used at the end of the ventilation device; not situated <600 mm from an opening in an "A" or "B" class division or "B" class ceiling; not more than 0.2 m^2 sectional area; and made of heat resisting noncombustible material (internally and externally faced with low flame-spread membranes having a calorific value 45 MJ/m^2 of their surface area for the thickness used. Exhaust ducts from galley ranges that pass through accommodation spaces or spaces containing combustible materials will now be required to have a fire damper in the upper end of the duct, in addition to the lower end. Exhaust ducts from galley ranges that pass through accommodation spaces or spaces containing combustible materials will now be required to have a fire damper in the upper end of the duct, in addition to the lower end.

Ref: SOLAS II-2, Reg 9 Ventialtion Ducts

Que. **Supplementary Lighting on Passenger Ships**
What is the Supplementary lighting requirement on new build passenger ships carrying >12 passengers built on or after 1st July 2010?

Ans. Supplementary lighting shall be provided in all cabins to clearly indicate the exit. The lighting is to be powered from an emergency source of power or have self-contained source of electrical power in each cabin. Lighting shall automatically illuminate when power to cabin lighting is lost and remain illuminated for at least 30 minutes.

Que. **Bunker Tank Protection**
What are the requirements for all new build ships constructed on or after the 1st August 2010 with respect to bunker tank protection?

Ans. Ships having an aggregate FO capacity of 600 m^3 and greater are required to "protectively locate" each bunker tank (which excludes tanks that do not normally carry fuel oil such as overflow tanks) having a capacity greater than 30 m^3 in accordance with the requirements of MARPOL Annex I, Reg 12A.

Ref: MARPOL Annex I, Reg 12A Bunker Tank Protection MEPC.141(54)

Que. **HCFCs**
Can HCFCs be used in air conditioning units?

Ans. This directive prohibits the use of virgin HCFC in the maintenance and servicing of refrigeration and air conditioning units on EU flagged ships.

Note:
The prohibition extends to non-EU flagged ships where an EU company is performing maintenance or servicing on board.

New or "virgin" HCFC cannot be supplied within the EU after 31 December 2009, but recycled HCFC may be available until 2015.

Ref: EU Directive – Use of HCFCs (2009/1005/EC)

Que. **Revised NO$_x$ Technical Code**
What are the Tier I NO$_x$ emission standards for engines and when do they need to be fitted?

Ans. When an upgrade kit has been approved for the relevant engine and becomes available in the market, engines (>5000 kW & 90 liters displacement) installed on ships will need to comply with the Tier I NO$_x$ emission standard (17.0 g/kWh when rpm <130; 45 n(-0.2) g/kWh when 130 n <2000 rpm; 9.8 g/kWh rpm 2000).

Ref: SOLAS II-1/3-9

4. Questions on Forthcoming Regulations Still to be Adopted or Enter Force

Que. **Freefall lifeboats**
SOLAS will introduce an amendment on 1st January 2014 that affects freefall lifeboats, what is contained in this amendment?

Ans. The operational testing of free-fall lifeboat release systems shall be performed either by free-fall launch with only the operating crew on board or by a simulated launching carried out based on guidelines developed by the Organization.
http://www.shippingregs.org/refdocs/MSC.325(90).pdf

Que. **Tankers**
Are tankers permitted to blend liquid cargoes at sea?

Ans. From 1st January 2014 a New regulation 5-2 is added to SOLAS Ch VI, Reg 5-2 on the 'Prohibition of the blending of bulk liquid cargoes and production processes during sea voyages'.
http://www.shippingregs.org/refdocs/MSC.325(90).pdf

Que. **New Chapter 4 to MARPOL Annex VI**
On 1 January 2013, a new Chapter 4 to MARPOL Annex VI will enter into force, do you know what this contains?

Ans. For new ships: Regulation 20 (Attained EEDI) of MARPOL Annex VI, as amended, requires that the Energy Efficiency Design Index shall be calculated taking into account the guidelines developed by the Organization. The 2012 Guidelines on the method of calculation of the attained Energy Efficiency Design Index (EEDI) for new ships, are set out at annex to the present resolution.
http://www.shippingregs.org/refdocs/MEPC_Res212.pdf

For existing ships: Regulation 22 of MARPOL Annex VI, as amended, requires each ship to keep on board a ship specific Ship Energy Efficiency Management Plan (SEEMP) taking into account guidelines developed by the Organization.
http://www.shippingregs.org/refdocs/MEPC_Res213.pdf

Que. What is EEDI?

Ans. The Energy Efficiency Design Index (EEDI) is a technical instrument that sets energy efficiency levels for new ships of >400 GT and some major conversions. It is intended to reduce the amount of CO_2 emissions from ships through improvements in energy efficiency.

EEDI is a measure of the volume of CO_2 produced by a ship per unit of transport provided (capacity), ie the ship's CO_2 efficiency. It is far from a simple regulatory formula.

Que. **EEDI**
Where would you find information on how to calculate the Energy Efficiency Design Index (EEDI) for New Ships?

Ans. '2012 Guidelines on the method of calculation of the attained Energy Efficiency Design Index (EEDI) for new ships' are set forth in the new regulation 20 of MARPOL Annex VI.

MARPOL Annex VI, Ch 4 (new chapter enters into force on 1 January 2013) http://www.shippingregs.org/refdocs/MEPC_Res212.pdf

Que. **SEEMP**
Where would you find information on the development of a Ship Energy Efficiency Management Plan (SEEMP) for existing ships?

Ans. '2012 Guidelines for the development of a Ship Energy Efficiency Management Plan (SEEMP)', are set forth in regulation 22 of MARPOL Annex VI.

MARPOL Annex VI, Ch 4 (new chapter enters into force on 1 January 2013), http://www.shippingregs.org/refdocs/MEPC_Res213.pdf

Que. **What ships does EEDI apply to?**

Ans. Under the new Chapter 4 to MARPOL Annex VI from 01/01/2013, an Energy Efficiency Design Index (EEDI) is required for new ships. EEDI is a non-prescriptive, performance-based mechanism that leaves the choice of technologies to use in a specific ship design to the industry. As long as the required energy-efficiency level is attained, ship designers and builders would be free to use the most cost-efficient solutions for the ship to comply with the regulations.

Que. **What is SEEMP?**

Ans. Under the new Chapter 4 to MARPOL Annex VI from 01/01/2013, the Ship Energy Efficiency Management Plan (SEEMP) is required for all ships.

SEEMP establishes a mechanism for operators to improve the energy efficiency of ships.

Que. **RO Code**
What is the RO Code?

Ans. The RO Code is the new *Code for Recognized Organizations* which was adopted at MSC 92 and take effect on 1st January 2015.

The Code aims to provide a standardised, global approach to assist flag States in recognising, authorising and monitoring their ROs. The new Code will apply to all ROs that are either being considered for recognition or are already performing statutory certification and other services for flag States.

Que. **IGF Code**
What is the IGF Code?

Ans. The International Code of Safety for Gas-Fuelled Ships (IGF Code) is the new code, still under development, that will provide industry standards and guidance for ships that use fuels with a flashpoint less than 60°C.

It is hoped that the draught of the document will be available for approval at the next Maritime and Safety Committee (MSC 93) in 2014, when a provisional entry into force date could then be agreed subject to Adoption and ratification.

Que. **Emergency Towing Arrangements – Cargo Ships >500 gt**
What are the emergency towing requirements that entered in to force for cargo ships on 1st January 2012?

Ans. By 1st January 2012, a procedure for establishing capabilities to tow the ship from the fore and aft locations had to be provided on board all cargo ships >500 t.

This procedure is to be carried onboard for use in emergency situations and shall be based on existing arrangements and equipment available onboard the ship taking into account MSC.1/Circ.1255.

Ref: SOLAS Ch II Reg 1/3-4 Emergency Towing Procedures MSC.256(84)

Note:
Emergency towing arrangements should be designed to facilitate salvage and emergency towing operations, primarily to reduce the risk of pollution. At least one of the arrangements should, at all times, be capable of rapid deployment in the absence of main power on the ship to be towed and provide easy connection to the towing vessel.

Such fittings, located at both ends, usually consist of an anchor point in the form of a 'Smit Bracket', to which is secured a chafing chain that is led through a panama fairlead. The purpose of this arrangement is to permit a tug to take the vessel in tow in the event that the vessel starts to founder.

The principal idea is that the ship's crew, prior to abandoning the vessel, will have the opportunity to deploy the chafing chains through the panama chocks, making them accessible to arriving salvage tugs.

Even with no-one left onboard, with weather permitting, a salvage tug would then have the ability to secure a towline to the anchored chafe chain.

Que. **Lifeboat Fall Preventer Devices**
What are Lifeboat fall preventer devices?

Ans. In 2009, the 86th session of IMO's Maritime Safety Committee approved Guidelines for the fitting and use of fall preventer devices (FPDs). It was emphasised that FPDs are only to be considered as an interim risk mitigation measure and are only to be used in connection with existing on-load release hooks, and that **wires or chains should not be used as FPDs** as they do not absorb shock loads.

Note:
FPDs are to be used at the discretion of the Master, pending the wide implementation of improved hook designs with enhanced safety features.

It was further recognised that a number of the current designs of on-load release hooks are designed to open under the lifeboat's own weight and often need to be held closed by the operating mechanism, with the result that any defects or faults in the operating mechanism, errors by the crew or incorrect resetting of the hook after being previously operated can result in premature release.

Ref: GUIDELINES FOR THE FITTING AND USE OF FALL PREVENTER DEVICES (FPDs) MSC.1/Circ.1327, 11 June 2009

In March 2012 the MCA issued MGN 445 (M+F)Lifeboats: Fitting of 'Fall Preventer Devices' to Reduce the Danger of Accidental On-load Hook Release.

Where they strongly urged that all UK vessels fitted with lifeboat on-load release systems should be equipped with fall preventer devices (FPD) pending the evaluation of the systems for compliance with the requirements of the revised LSA Code.

Note:
This notice is only applicable to davit-launched lifeboats fitted with on-load release hooks. FPDs should be fitted in accordance with MSC.1/Circ.1327 'Guidelines for the fitting and use of fall preventer devices (FPDs)'.

Que. **Replacement of Lifeboat On-Load Release Mechanisms**
What are the SOLAS amendments that affect lifeboat release mechanisms?

Ans. The Maritime Safety Committee of the IMO has adopted a new paragraph 5 of SOLAS regulation III/1 to require lifeboat on-load release mechanisms not complying with new International Life-Saving Appliances (LSA) Code requirements to be replaced no later than the first scheduled dry-docking of the ship after 1 July 2014 but, in any case, not later than 1 July 2019.

The SOLAS amendment, which is expected to enter into force on 1 January 2013, is intended to establish new, stricter, safety standards for lifeboat release and retrieval systems, aimed at preventing accidents during lifeboat launching, and will require the assessment and possible replacement of a large number of lifeboat release hooks. http://www.shippingregs.org/refdocs/MSC_Res317.pdf.

Que. **Lifeboat Release Hooks not Meeting the Requirements**
Can you give examples of lifeboat release hooks which may not meet the relevant requirements, particularly, if they fall out of tolerance, due to wear?

Ans. Flat to flat cam hooks
Forward rotating round cam with a self-locking capability
Flat to flat cam with some self-locking capability.

Que. **Lifeboat Hooks that Fail to Comply**
If a design review of lifeboat hooks reveals that a release mechanism does not comply with paragraphs 4.4.7.6.3 to 4.4.7.6.5 of the LSA Code, or a design review cannot be carried out because design documentation is not available what action should be taken?

Ans. Release mechanisms that do not comply should be replaced at the earliest available opportunity, and no later than the next scheduled dry-docking after the entry into force date.

Until the release mechanisms are replaced, additional safety measures, including use of fall prevention devices in accordance with MSC.1/Circ.1327, should be employed.

Note:
As an alternative to replacement, the hooks may be modified to comply with the requirements of paragraphs 4.4.7.6.3 to 4.4.7.6.5 of the LSA Code, as amended by the appropriate resolution at MSC.88, provided that the modifications are approved by the Administration.

Que. **Revised NO_x Technical Code**
What are the Tier II NO_x emission standards for engines and when do they need to be fitted?

Ans. Diesel engines (>130 kW) installed on ships are to meet the Tier II NO_x emission standard (14.0 g/kWh when rpm <130; 44 n (-0.2) g/kWh when 130 n <2000 rpm; 7.7 g/kWh rpm 2000) by 01.07.2011.

Que. **Bunker Fuel – Sulphur Levels**
What was the sulphur limit in bunker fuels restricted to from 1st January 2012?

Ans. A reduction in sulphur limit to 3.5% globally for bunker fuels.

Ref: MARPOL Annex VI – Ch 3 Reg 14

Que. **Enclosed Spaces**
When not in use, in what condition should an entry door or hatch to an enclosed space be left?

Ans. Entry doors or hatches leading to enclosed spaces should at all times be secured against entry, when entry is not required.

Ref: 'Revised Recommendations for entering enclosed spaces aboard ships' were approved, at the 27th IMO Assembly 21-30 November 2011.